Tawona Honour Matenda, T. Amtaita, M. Mrema

An assessment of the nursery industry in Manicaland, Zimbabwe in 2008

Tawona Honour Matenda, T. Amtaita, M. Mrema

An assessment of the nursery industry in Manicaland, Zimbabwe in 2008

GRIN Verlag

Bibliografische Information der Deutschen Nationalbibliothek: Die Deutsche Bibliothek
verzeichnet diese Publikation in der Deutschen Nationalbibliografie; detaillierte bibliografische Daten sind im Internet über http://dnb.d-nb.de/ abrufbar.

1. Auflage 2010
Copyright © 2010 GRIN Verlag
http://www.grin.com/
Druck und Bindung: Books on Demand GmbH, Norderstedt Germany
ISBN 978-3-656-07295-9

AN ASSESSMENT OF THE NURSERY INDUSTRY IN MANICALAND, ZIMBABWE IN 2008

Tawona MATENDA[1], T.A MTAITA[2] and M. MREMA[2]

Africa University, Faculty of Agriculture and Natural Resource, P.O Box 1320,Mutare, Zimbabwe

---

ABSTRACT

Agricultural nurseries are key in the production of horticultural and forestry seedlings. Various nursery media is used to raise these seedlings. The study was aimed at determining the researchers' views and opinions on the nursery practices that the farmers employ. The survey was carried out to establish the productivity and types of media used in the nursery industry within Manicaland, Zimbabwe. In this study, the survey was conducted at provincial level within five (5) selected districts of Manicaland Province. In each district, all the existing nurseries were selected. Data was analyzed using the descriptive analysis. Nursery holders were classified as entrepreneurs, farmers, local government owned and non –profit making organizations owned. Farmer nurseries were the largest category comprising of 56.4% of the total sampled population. Most of the seedlings were raised in containers with the exception of vegetable seedlings which were raised in seed beds. Nursery media comprising of top soil and compost proved to be most popular amongst nursery owners. Pine media used proved to be least used due to its low water holding capacity and lack of proper extension training on its use.

INTRODUCTION

Investment in forest tree planting is increasing annually. FAO surveys (1993) indicated that forest plantations in the tropics alone expanded from a total of 18 million hectares in 1980 to 44 million hectares in 1990. In tropical and subtropical countries, the majority of trees planted are raised in containers. Nursery research has demonstrated that development of a fibrous root system is essential for good quality seedlings. Root absorptive efficiency is directly related to its surface area and fibrous roots provide greatest surface area. Development of fibrous roots is related to the porosity of the potting mixture which in turn is related to the organic fraction of the mixture. Pine bark is a light weight soiless media which is easy to transport. Organic substrates provide adequate nutrients to the seedlings, better root substrate relation than conventional soil mix and less pre-dispose the seedlings to soil borne pests and diseases (Adams et al., 2003; Akanbi et al., 2002).

The major aims of this work were (1) to assess the types of substrates used in the production of seedlings in five (5) districts in Manicaland and (2) to assess the operations of selected extension and registered nurseries in Manicaland.

MATERIALS AND METHODS

In this study, the survey was conducted at provincial level within selected districts of Manicaland Province. The study was aimed to determine the researchers' views and opinions on the nursery practices that they employ. The study focused on the assessment of different types of nursery media used within selected nurseries and assessment of productivity of extension and registered nurseries. Manicaland Province forms the eastern border of the country. It has an area of 36,459 km² and a population of approximately 1.6 million (2002 census). Mutare is the capital of the province. The province is subdivided into seven districts namely Buhera, Chimanimani, Chipinge, Makoni, Mutare, Mutasa and Nyanga. Nurseries located in Manicaland were considered in the survey. The Forestry Commission Department of Conservation and Extension was selected for informant interviewing because it had initiated extension nurseries within the province so as to encourage afforestation and community awareness in conservation of forests. From the informant interview conducted with personnel

from the Forestry Commission Department of Conservation and Extension on the list of nurseries in the province, five districts were chosen because they conduct horticultural nursery culture either as individual initiative or in collaboration with department. From each district, all existing nurseries were considered for study. The number of nurseries in each district (Table 1) varied depending on the individuals working on nursery culture.

Table 1: Number of nurseries per district

District	Number of nurseries
Chipinge	4
Makoni	6
Mutare Urban	4
Mutasa	4
Nyanga	3

Each district selected was composed of various types of nursery holders. These can be categorized as entrepreneurs, farmers, local government and non–profit making organizations. Entrepreneurs were those who embark on nursery culture for the sole purpose of obtaining income. They could either buy seedlings for further sale from individuals who raise seedlings on their own landholdings or raise their own seedlings or combine both options so as to meet the market demands

Secondary data was used to obtain recorded data on nurseries. Maps showing the location of districts were obtained from the Ministry of Lands and Rural Development. Data pertaining to the Rural Afforestation Project was obtained from Forestry Commission Department of Conservation and Extension library. An informant interviewing was conducted to gather information pertaining to extension nurseries that were initiated by the department with the aim of promoting afforestation and forestry sustenance in the communal areas. Information obtained from the key informant interviews revealed that individual farmers, prison services and schools were working together with the Forestry Commission's Department of Conservation and Extension to start nurseries.

A questionnaire was designed and administered by the researchers. Visual assessment of the nursery site was also done to make sure that the information that was provided in the survey complied with facts on the ground. Owners of the nurseries comprised individual farmers, entrepreneurs, schools and municipal councils.

A quantitative analysis of the nursery questionnaire was carried out using the Statistical Package for Social Sciences (SPSS) Release 11.0. Data was also analyzed using descriptive statistics.

RESULTS

Nursery ownership and land holding

The study showed that entrepreneurs formed 14.3% of the total nursery owners. All the farmers interviewed were working together with Forestry Commission Conservation and Extension personnel. The study found that all the raised seedlings were sold to individuals or other entrepreneurs for resale in urban centres or used within the community to foster forestry development through the reafforestation programmes. All the proceeds went to the farmer. The highest percentages of nursery holders were found to be farmers (56.4%). Local government nurseries are those in the custody of the city or town council. These were the least in number at 4.8%. Local government nurseries were found to be located in Mutare urban district only. They provided seedlings to the residence of the city. Non–profit making organizations comprised of non–governmental organizations (NGO), schools, colleges and prisons. Their main aim for nursery production was to provide seedlings to the community so as to improve community livelihood and dissemination of knowledge. These formed 28.6 % of the total nursery owners.

Table 2 shows the characteristics of the interviewed nursery holdings in Chipinge, Makoni, Mutare Urban, Mutasa and Nyanga, districts. Makoni district had the highest percentage of nursery holdings (28.6%) and all these are still under the supervision of Forestry Commission Department of Conservation and Extension. Nyanga district had the lowest number of nursery ownership (14.3%). Chipinge, Makoni and Mutare Urban had the same number of nursery ownership (19.0%).

The average number of nurseries per district was 1.90 (standard deviation (sd): 1.338). Farmer nurseries within the different districts were found to be all working on nurseries as an initiative from the Forestry Commission Department of Conservation and Extension. The nurseries were set-up between 1954 and 2007. 90.5% of the land the nursery belonged to owners of the land. Only 9.5% of the nurserymen were using rented space for their activities. Those who rented had to pay rent ranging between USD 40.00 and USD 50.00 for approximately $200m^2$ to $300m^2$.

Table 2: Type of nursery ownership in Manicaland

Location of nursery	% nurseries per district
District	
Chipinge	19.0
Makoni	28.6
Mutare urban	19.0
Mutasa	19.0
Nyanga	14.4
Type of land ownership	
Owned	90.5
Rented	9.5

Ownership by sex

Women were found to play a big role in nursery production. Their role varied with each enterprise. The survey showed that 14.3% of the nurseries interviewed were owned by women. Chipinge (25%), Mutasa (25%) and Makoni (33.3%) districts had women claiming ownership of the nurseries at 25%, 25% and 333% respectively.

Female ownership came as a result of the male absence as heads of the families in the lives of the women. Death of a spouse and emigration of the spouse to the city were the reasons given by the women for spouse absence. The women had to embark on nursery production so that they can obtain additional income to sustain the families.

85.7% of the nursery enterprises were owned by males or the organizations headed by males with some women participating in the production activities of nursery products. Mutare urban and Nyanga had no female nursery owners. As shown in Figure 1, 19% of the women were assistants in the running of the nurseries. Their duties included maintaining the nursery seedling and propagation of materials for use at the nursery. Proceeds obtained from the sale of the produce would go to the owner of the nursery enterprise who in most cases is male.

The greatest number of enterprises had non participation of women in nursery production (67%) (Figure 1). This was due to some of the nursery owners being entrepreneurs and therefore operated their businesses independent of women involvement. City council workers comprised of women working directly with the products whilst the males were responsible for ferrying produce from one point to the other or the supplying of the media.

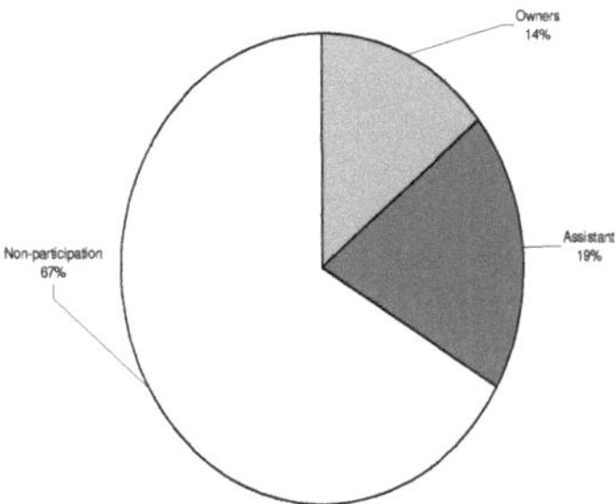

Figure 1: Role of women in nursery production

Production area

The area under production varied according to the location, availability of inputs for production and market demand. Production area comprised of space to be used for preparation of the media, raising the seedlings and preparation of the seedlings for market. Most of the area under production was below $100m^2$ (standard deviation (sd): 1.63). Table 3, shows that 42.9% of the nurseries have a total production area of less than $100m^2$. These constituted entrepreneurs, farmers and non-profit organizations (Figure 2). Turnover period of time of seedling grown and sold varied with each nursery. During the hot wet season there was a high turnover rate as people would be establishing orchards. The hot dry season was characterized with a low turnover rate. It was observed that though a large

portion of land was allocated to nursery production, not all of the land was serving this purpose. Approximately 35% of the land was lying fallow. Container areas were more seedlings under production per unit land than the rest of the land allocated to production.

Table 3: Production area under nursery culture

Characteristic	Percentage of owners of nurseries
Area under production	
Total production area	
$< 100m^2$	42.9
$100 - 250m^2$	19.0
$300 - 500m^2$	14.3
$600 - 750m^2$	0
$> 800m^2$	23.8
Container area	
$0\ m^2$	9.5
$< 50m^2$	42.9
$50 - 150m^2$	23.8
$200 - 300m^2$	4.8
$400 - 500m^2$	14.3
$> 600m^2$	9.5

Farmer nurseries had the highest total land area in the less than $100m^2$ group(Figure 2) which was allocated to production as compared to the other groups. As indicated in Table 4.0 shows that 23.8% of the nurserymen had a total production area of more than $800m^2$. These constituted farmers, local government and non-profit organizations (Figure 2). The study revealed that, all grown nursery species (Table 7) were raised in containers with the exception of vegetable seedlings which were raised in seed beds.

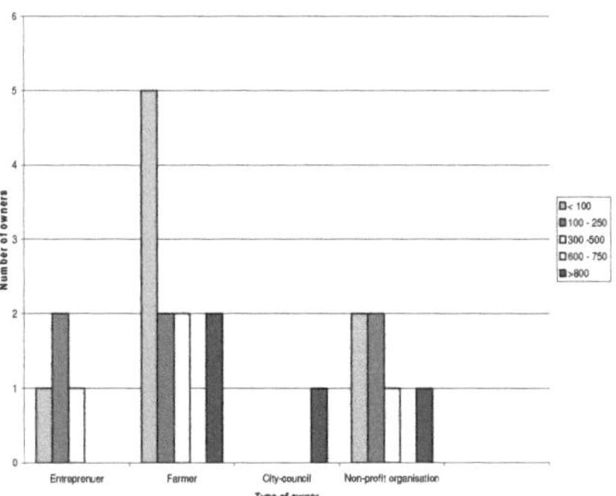

Figure 2: Distribution of total nursery area

Container area at two non-profit organizations exceeded $600m^2$. City council and farmer nurseries had container nursery area of less than $50m^2$. Although the city council has a large area allocated to nursery production ($>800m^2$), it had a smaller proportion allocated to container production and the rest of the area had to be used for other activities such preparation of propagation materials and operations such as grafting and budding. Container area for entrepreneurs ranged from $50 - 300m^2$ (Figure 3).

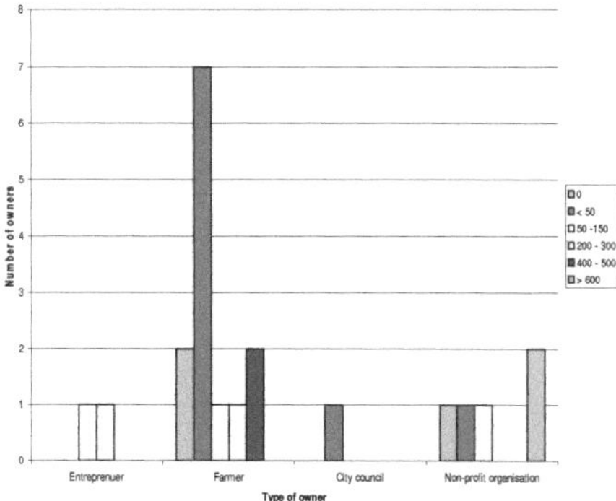

Figure 3: Distribution of container nursery area

Opinion on nursery expansion

Nursery owners had varying opinions on whether they wanted to expand on the nursery area or not. Owners who wanted to expand on the area were 42.9% (standard deviation (sd): 0.507) with the rest considering non expansion

(57.1). Table 4 shows that 45% were satisfied with the land allocated to them and wanted to fully utilize the land so as to maximize on production (standard deviation: 1.67).The nursery owners gave various reasons regarding the decision they had made on the possibility of expanding on the nursery area as given below.

Farmers'

Farmer nurseries (100%) located in the communal areas were content with the area for nursery production. The land under production was as a result of inheritance in the family and there was no possibility of more land for nursery. The farmers did express intention on expanding through relocation of the nursery business to a site which is located near a business centre or town. Relocation would ensure a larger clientele base and therefore an increase in the amount of income.

City council

Mutare city council officials did not have intentions of expanding on nursery area because production at the nursery site was adequate to meet their customer requirements.

Non-profit organization

In Mutasa district, partnership of Mutasa Primary School with Environment Africa was highlighted as the reason for the need of expansion of the current nursery area (Table 4). The school was shifting from the fruit tree and eucalyptus seedling production to herbal seedlings production and permaculture. In Makoni district, Non-governmental organizations expressed the need to expand on the nursery area since there was a growing demand in the herb seedlings. Demand has been raised by both individuals and communities that they work in promoting of herbal gardens. Herbal gardens promotion was being done to try and supplement on modern medication.

Entrepreneur

Nursery holders who rented the nursery area sited the need to expand on the area under production. Due to the growing demand for tree seedlings, the area currently used was proving to be a limiting factor (40%). Space was not only used for placement of the marketable products but also used for preparation of the media and raising of seedlings until they are of marketing quality. Entrepreneurs in Nyanga district expressed the need for local government to allocate them more land so that they might be able to expand their business ventures. The entrepreneurs also sited water shortages as a disadvantage to land area expansion (Table 8). In general, the water delivery system in the residential areas is very poor with water cuts occurring often and for several hours at a time. This situation has resulted in the owners cutting down on production so that they are able to better manage the seedlings with the resources available. An increase in the demand for the seedlings especially fruit tree seedlings was sited as a reason given by the owners for wanting the expansion of the current nursery area if the water situation was resolved.

Table 4: Reasons for expansion/non-expansion of nursery site

Reasons For expansion	Percentage
Partnerships with other organizations	5
Increase in demand	30
Inadequate land	40
Non-expansion	
Full utilization	45
Economic	10
Lack of resources	15

Nursery media used and transportation

The type of media used was dependent on the accessibility of the media and the extension media methods that were disseminated by the extension workers. Locally available media were used by the nursery holders. The most used type of media was a combination of topsoil and compost only.

Table 5: Types of media formulation used by nurseries

Type of media used	Percentage farmer use	Standard deviation
Top soil & compost	42.9	0.782
Pine, topsoil & compost	23.8	1.789
Sand and compost	33.33	1.069

According to the survey, 66.7% (Table 5) of the nurseries used top soil and compost media

either as the sole media or in combination with pine bark. Compost was used for nutrient enhancement and better media aggregation. Pine bark was used together with other media methods by 23.8% of the nursery owners.

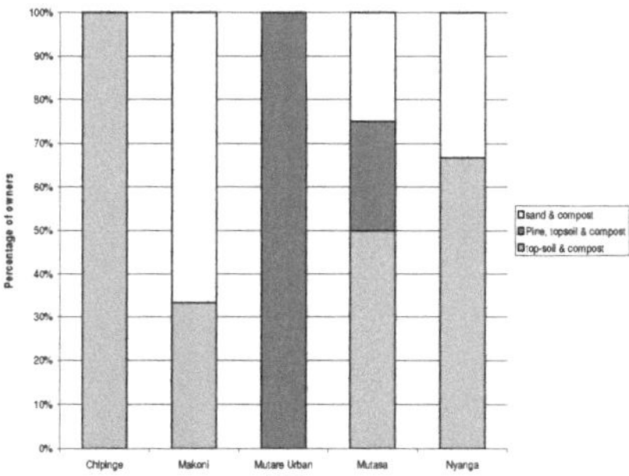

Figure 4: Types of media formulations used by nurseries

Entrepreneurs and non profit organizations surveyed used this combination (Figure 4) because these nurseries are located near pine bark source. For example, pine sawmills are located in the city of Mutare and the pine can be obtained for free or for a small fee of USD 0.67 at the sawmills. Farmer nurseries did not use pine bark combinations (Figure 4) because of lack of capital to finance the transportation of the pine bark from the mill to their operational farms. One farmer in Makoni district discredited the use of pine bark as a media because the media is of low water holding capacity. Use of pine bark mixtures proved to be laborious as the seedlings would have to be watered more frequently.

Most of the nursery owners who used this media, owned the transportation that was used to ferry the media from the source to the nursery site (90%). Only 10% of the owners had to hire transportation (Table 6). These had to incur a cost ranging from USD 167.00 to USD 500.00 to ferry the media. Fuel costs were incurred by the owners who had tractors and trucks as the means of transportation of the media. Fuel was costed per trip at an average USD2.00 per kilometer made from the source of media to the nursery site. On average, entrepreneurs who used trucks to ferry the media had to hire the truck, 3 times per month. Farmer nurseries that used 50kg

bags to carry the media had to acquire more than 500kg of the media of approximately $300m^2$. None of the nurseries surveyed had access to government subsidized fuel therefore they had to rely on the informal market that charge US$1.20 per litre.

Table 6: Ownership of transportation

Ownership of transport	Percentage
Hired	10
Owned	90

Different types of transportation means were used by the nursery owners depending on the availability of the resource such as media and capital and the frequency of media used per production season. Type of vehicles for transportation ranged from trucks, tractors, ox drawn and wheelbarrows (Figure 5). Farmer nurseries usually relied on the use of the wheelbarrow as a means for media transportation. This method of transportation proved to be effective since ownership is at household level and the source of media was within 500m from the nursery site in most cases. For distances further than 500m the nursery holders had to use trucks and tractors. Entrepreneurs, city council and non-profit organizations used tractors and trucks to ferry the media (Figure 5). On average entrepreneurs would use the truck to ferry the media three times a week whilst four tractor trips were required per month by the city council for more than $500m^2$ of production area. However not all of the media acquired was used by the nursery holders. About one third of the total media acquired was observed still in heaps in the nursery media preparatory section.

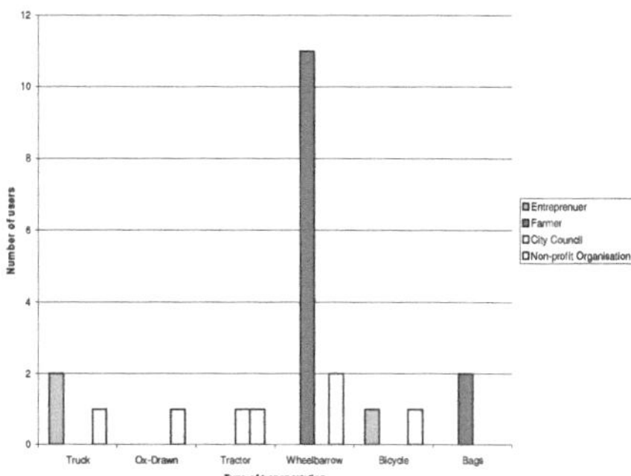

Figure 5: Transportation used by the nursery owners.

Nursery training

The study found that in order to ensure the successful raising of nursery seedlings, the nursery holders had to learn the art of grafting, budding and cutting back. The nursery holders interviewed admitted to receiving training on how to tend to the nursery. Training was attained either from extension workers (74.8%) or informal learning through experienced personnel already in the nursery business (45.2%).

The Department of Conservation and Extension has provided one extension worker per district. The extension officers were provided with transportation to carry out farm visits once a month but due to the current change to economic hardships (Annual Inflation rate 1021%), the department has been unable to provide transportation. Nurseries within the vicinity of the extension offices attained visitation once a month while those located more than 80km away received visitation once in three months. Nursery owners acknowledged that they had not received training concerning the managerial aspects of the nursery. No proper records were kept concerning the day to day running of the nurseries. Each nursery holders relied on his/her memory. School nurseries were used as a platform to educate students on the importance of tree planting and management techniques in order to help overcome environmental problems such as deforestation and soil erosion which are becoming a serious problem as population increases. Excess seedlings were given to students who inturn were encouraged to plant them on their parents' plots or farms. This would provide secondary training i.e. student to parent training on the importance of reafforestation.

Species grown

A variety of species are grown in the nurseries according to the location and objective of the nursery enterprise. All the holdings planted a variety of seedlings ranging from herbs, flowers, fruit tree and propagation materials. Over 98% of the herbs were grown by non-profit making organizations (Table 7) because non-governmental organization headed nurseries promotes herbal gardens in homesteads. The centre acts as a source of herbal seedlings for the local community. Fruit trees and evergreen trees were sold amongst all the nursery holdings with the greatest percentage (77.1) of fruit trees being grown by farmers because the province is conducive for the production of a variety of tree species since it encompass all

The five agro-climatic zones. Fruit trees that proved most popular are oranges, peaches, lemon, nartijies, plum, apricots, and grape.

Evergreen trees such as the eucalyptus were mostly grown and sold by the farmer nursery owners.

There has been diversification in the types of seedling species that are grown by the nurseries. Vegetable seedlings were not popular among the nurseries with only 14.4 % of the owners embarking in production. Quantities produced ranged from 200 to 1000 seedlings annually.

Production constraints

Constraints encountered by the nurseries ranged from environmental, managerial and local government failure to provide services. Severity of constraints varied among the owners. This could be attributed to the location of the owners. Just before the survey the department of conservation and extension had been providing the farmer nurseries with packaging materials, but this ceased due to lack of funding in the department. Farmers now had to buy their own packaging materials. Despite the failure by department of conservation and extension to provide packaging materials, 89.7% of the nursery holders sited packaging not as a constraint to production (Table 8). Some of the farmers interviewed had resorted to using polythene

waste such as empty fruit juice containers to use as media containers

Table 7: Species grown by the nursery holding

Types grown	% Nursery holding			
	Entrepreneur	**Farmer**	**City council**	**Non-profit making organizations**
Deciduous shade and flowering trees	49.0	0	49	2
Deciduous shrubs	48.5	0	48.5	2.9
Evergreen shrubs	42.6	0	28.4	29.0
Evergreen trees	17.2	74	8.6	0.2
Vines and ground covers	38.2	5.1	25.5	31.2
Flowering annuals	18.4	0.6	80.6	0.3
Herbs	0.7	0.1	0.8	98.5
Flowering potted plants	73.1	1.7	17.4	7.7
Fruit trees	17.6	77.1	3.9	1.4
Foliage	91.7	0	8.3	0
Propagation materials	16.7	0	83.3	0
Vegetable	58.8	11.8	29.4	0

Table 8 : Constraints associated with nurseries

Constraint	% of nursery owners	
	Not severe	**Severe**
Water	47.3	56.6
Transportation	79	21
Land availability	79	21
Diseases	56.7	47.3
Distance from media source	84.2	15.8
Continuity of media supply	100	0
Continuity of seedling supply	100	0
Labour	79	21
Weather uncertainty	73.6	26.4
Marketing	84.2	15.8
Competition	100	0
Managerial expertise	73.6	15.8
Packaging	89.5	10.5
Natural disasters	94.7	5.3
Other production materials	84.2	15.8

Farmers located in the Nyanga area sited water and land availability as the major constraints to production (56.6%) (Table 8). They attributed this problem to failure of the local government to supply adequate water to the community as a result they were being forced to scale down on production. The local government was delaying in the allocation of land and thus the owners had to depend on the land around their homesteads for production. This was also not enough as they were getting orders for seedlings far beyond their production capacity.

All the nursery owners interviewed acknowledged that continuity of media and seedling supply was not a constraint in the running of the nursery enterprise (Table 8). The media could be obtained within reasonable distance from the nursery site and was either for free or of affordable cost. The nursery sites which encountered constraints

associated with the marketing of their produce are those located away from the business centres (15.8%). The researchers established that there is a high demand for fruit tree seedlings in all the nurseries surveyed however the demand was seasonal, that is, mostly during the hot wet season. The nursery business is not a business that is popular among locals in all the districts that were surveyed. The nursery owners face very little competition from each other. Nursery owners in Nyanga and Chipinge district occasionally come together to discuss ways of improving their business and those who have been in the business for a longer time offer assistance to those who are struggling. Assistance can be in the form of training on raising of the seedlings and in some cases they refer customers to their colleagues if they know they have a specie the customer wants.

As shown in Table 8, 47.3%, nursery holders cited diseases as a constraint to production. Damping–off and blights were some of the diseases that mostly affected their seedlings. The nursery holders had to resort to discarding of the affected seedling through burning or dumping them. Diseases were found to be most prevalent during the rainy season.

Nursery owners did not cite labour as a constraint as nursery production is done at a small scale. Labour obtained from family was adequate.

DISCUSSION

In a similar study, Mvududu (1992), also showed that nurseries in Manicaland were owned by individuals, schools, farmer group, armed forces and forestry commission. There has been an increase and diversification of the ownership of the nurseries from the period of 1992 to 2007. Diversification has been brought about by the increase in non governmental organizations assisting in the country and efforts by the government to encourage entrepreneurship among individuals.

A survey by Mvududu (1992) in Manicaland revealed that a total of four (4) nurseries were handed over by Forestry Commission to farmer groups. The results of this study indicate that there has been an increase in the number of nurseries found within the district and also nurseries being established in districts which were initially not targeted by the afforestation program of Forestry Commission such as Chimanimani, Nyanga and Mutasa. Producing seedlings in containers would ensure easier handling of the seedling when they are sold. The shift from fruit tree production and eucalyptus seedling production to herbal seedlings is because of the partnerships that are formed with other organizations. Forestry Commission encourages the production of fruit trees and eucalyptus seedlings so that the local community can assist in reafforestation of their surrounding area and also for sale so that the project can be self sustaining. Environment Africa NGO facilitates the implementation of sustainable livelihoods, schools environmental education programs, dialogue with industry for corporate social responsibility, and environmental rights (Environment Africa, 2008). The major problem entrepreneurs in Nyanga encounter is failure of local government to recognize nursery production as a business, thus they did not benefit when land for development became available. Pine is viewed as a locally available resource which can be obtained from sawmills located in the city of Mutare and other areas in Nyanga where plantation forestry is centered. Limited use of pine bark as a nursery media can be attributed to the insufficient training from extension workers that they receive and the lack of milled pine resources in the community. Though the extension workers educate them on various methods of media, they mainly promote the use of readily available resources which are within reach in the community. At the inception of the Rural Afforestation Project, the hierarchical structure of extension support ran from the provincial forest officers to District Forest Officers to Forest Assistants. The Forestry Commission extension workers would operate in close collaboration with AGRITEX (now known as AREX) and Ministry of Education staff. Forestry Commission extension workers carry out regular visits in farmer nurseries to ensure that the farmers are well educated about raising the seedlings and supplying any of the inputs that the department can afford. This could be due to the re-afforestation programme that was launched by the Forestry Commission Conservation and Extension department and the need to provide fuel wood nearer to homesteads. During the inception of

the afforestation programme (Mvududu, 1992), the seedlings produced comprised of eucalyptus, indigenous and exotic fruits. The diversity in seedling types could be attributed to the demand for diversity within the market.

Proper records pertaining to the expenses incurred and the sales realized are necessary in evaluating the viability and sustainability of nurseries. It is essential to measure the productivity of the enterprise so as to ensure viability. The extension department should embark on business training for nursery owners so that the owners of the business can ascertain the sustainability of the nursery enterprise.

Conclusion

In Manicaland, most of the nursery owners do not use pine bark as a nursery media for seedling growth due to its poor water holding capacity and poor information dissemination by extension workers on various mixtures used as media in nursery culture. Instead they opt to use top soil and manure which has its own limitations as a nursery media.

Reference:

Adams BA, Osikabor B, Abiola JK, Jayeoba OJ, Abiola IO 2003. Effect of Different Growing Media on the Growth of *Dieffenbachia maculata*. In The Role of Horticulture in Economic Development of Nigeria

Akanbi BW, Togun AO, Baiyewu RA 2002. Suitability of plant Residue compost as Nursery medium for some tropical fruit tree seedlings. Moor J. Agric. Res. 3: 24-29.

Enviroment Africa 2008. www.enviroafrica.org Accessed 5 March 2008

Mvududu S., 1992. Evaluation of the handed over nurseries. Forestry Commission. Unpublished. Pg. 28-42